BIRD FINDER

A Guide to Common Birds in Eastern North America

ROGER J. LEDERER
illustrated by ROGER C. FRANKE and CAROL E. BURR

Nature Study
Guild Publishers
an imprint of AdventureKEEN

HOW MUCH CAN A POCKET-SIZE BOOK DO?

This book will help the novice bird-watcher identify 59 of the most common species of birds east of the Mississippi River. To identify all 650 or so species you will need a more comprehensive field guide such as:

The Sibley Field Guide to Birds of Western North America, by David Allen Sibley. Alfred A Knopf, 2016.

Peterson Field Guide to Birds of Eastern and Central North America, Seventh Edition (Peterson Field Guides), by Roger Tory Peterson. Mariner Books, 2020.

For general information about birds of North America, we recommend:

National Audubon Society Birds of North America (National Audubon Society Complete Guides), 2021.

AMNH Birds of North America. American Museum of Natural History. A Photographic Guide. DK, 2020.

© 2024 Roger J. Lederer (text), Roger C. Franke and Carol E. Burr (illustrations); © 1990 Nature Study Guild • ISBN 978-0-912550-36-7 • Printed in China • LCCN 2024007748 • naturestudy.com

HOW TO USE THIS BOOK

Like comprehensive field guides, *Bird Finder* is organized in taxonomic order, beginning with grebes and ending with songbirds. There is no one way to identify birds, so when first using the book, you will have to page through it to identify the bird you are looking at. Once familiar with the book's organization, you will be able to go to the waterfowl or hawk descriptions quickly.

Because there are about 650 species of birds in the eastern United States, bird-watching could become overwhelming. But you have to start somewhere, and a simple book like this is a good choice.

Each page has a sketch of the bird and gives its common and scientific names; body length from beak to tip of tail (L) and wingspread (W); some identifying features such as eye stripes, wing bars, tail pattern, and behavior; and icons that indicate its usual habitat.

Because the birds in this book are among the most common birds you will see on a bird walk, they should be easy to find and identify. As you develop your skills as a bird watcher you will be able to identify other birds and move on to a comprehensive field guide.

Pied-billed Grebe
Podilymbus podiceps

Riding low in the water, the small chunky grebe has a brownish body and a white bill with a black patch, hence the pied (patched) name. Dives for invertebrates, frogs, and fish; eats its own feathers to protect its stomach from bones. Builds a floating nest of reeds in emergent aquatic vegetation. Babies ride on parents' backs after hatching. Grebes are excellent swimmers and divers but rarely fly, except to migrate. Swimming is aided by lobed toes and legs positioned far back on the body. Able to adjust buoyancy by trapping air between the body feathers to float and compressing feathers to sink.

L: 14"   W: 22"

Freshwater

Marshes

EGRETS
Great Egret
Ardea alba

Three white egrets are common over most of the US. The tall Great Egret has yellow legs, feet, and bill; the shorter **Snowy Egret** (*Egreta thula*; L: 24", W: 41") has a black bill and legs with yellow feet. Almost exterminated in the early 1900s for their plumes to decorate hats and clothing, the "feather trade" led to the formation of the Audubon Society. The **Cattle Egret** (*Bubulcus ibis*; L: 20", W: 36") has yellow legs and bill and, during breeding, yellow plumes on its head and back. Native to Africa, Cattle Egrets have spread across the US since their introduction in 1877. Like herons, egrets silently stalk their aquatic prey in shallow water.

L: 39" W: 51"

Freshwater   Marshes

Great Blue Heron
Ardea herodias

Widespread throughout North America, this heron is all gray with a white face and dark crown with short plumes. Like egrets, herons pursue fish, crayfish, snakes, and other animals near or in shallow water. Great Blue Herons quietly stalk their prey with slow, silent footsteps, then plunge their bill into the water to capture it. Although it's a large bird at nearly 4 feet tall, its hollow bones allow it to tip the scale at only 5–6 pounds. Both herons and egrets nest in large colonies.

There is no biological difference in herons and egrets; the names just come from different languages.

L: 46"   W: 72"

American Bittern
Botaurus lentiginosus

Streaked brown and white to blend in with the emergent plants of their watery environment. It's easy to unknowingly approach the stock-still bittern and then be surprised as it explodes into the air with a booming, bullfrog-like call. Standing with its bill nearly vertical, the bird has a 360-degree view of potential prey and predators, so it can react quickly. Primarily feeds on fish, as well as frogs, tadpoles, insects, crayfish, crabs, salamanders, snakes, and even flying dragonflies. Breeds over most of North America and migrates to the southern states and Mexico in winter.

> Herons, egrets, and bitterns stand quietly or wade slowly until they can stab or snatch a passing fish, frog, or crayfish. Some species spread their wings to attract aquatic prey to the shade or drop pieces of vegetation to bait small fish. Fishing is aided by their polarized eyesight.

L: 28"　W: 42"

Marshes

Canada Goose
Branta canadensis

Found throughout the US and common in winter. Wrongly called Canadian Geese (all North American geese nest in Canada), the Canada Goose has become two species: the larger Canada Goose and the much smaller **Cackling Goose** (*B. hutchinsii*; L: 25", W: 42"), which is found in the mid-Plains states and the East Coast in winter. There are also several races of the Canada Goose, such as dusky, lesser, and Aleutian, that differ slightly in size and coloration. They graze on grass, grains, and aquatic invertebrates. Lays 5–6 eggs in nest on ground, on a natural or artificial platform, and even in trees. Identified by its gray body, black neck, white throat patch, and raspy honk.

L: 40"   W: 60"

Snow Goose
Anser caerulescens

The Snow Goose comes in two versions: the white goose and the blue goose, which is mottled but mostly gray. Both have black wingtips. The all-white plumage serves as camouflage when nesting in snowy fields, while the blue goose coloration serves better after the thaw. Both breed, and interbreed, in the arctic region; migrate together through midwestern and coastal states; and winter along parts of the US East Coast. Feeds on grains, grasses, and aquatic vegetation. They nest from May through August and spend the rest of the time migrating south, then, when the time comes, north again.

L: 30"   W: 59"

Marshes

Wood Duck
Aix sponsa

One of the most attractive ducks, the male has a striking multicolored head, a greenish-purple back, red eyes, flowing green nape feathers edged by white, and a white neck and chin. The female, like most female ducks, is a dull brown. Unlike most ducks, Wood Ducks nest near water in tree cavities or artificial nest boxes. Adults fly into the nest without perching. The 10–15 fluffy, precocial young jump from the tree, sometimes dozens of feet high, and tumble to the ground. Feed on vegetation on surface of water or slightly below by dabbling—sticking their head down and rear up. Like other dabblers, they strain food from the water with their ridged bills. Live year-round in the Southeast and breed throughout the eastern US.

L: 19"   W: 30"

Mallard
Anas platyrhynchos

Throughout the northern hemisphere, the Mallard is probably the most familiar of all ducks, with the male's green head, white neck ring, cinnamon chest, and curly tail feathers; the female is a drab streaked brown. Its familiar quack is the classic duck sound. Taking off almost vertically, Mallards display their white tail feathers. They feed on aquatic vegetation or invertebrates. The female makes a feather-lined nest in grass and lays 8–15 eggs, which she incubates while the male is elsewhere undergoing a molt. Eggs hatch about a month later, with the female providing all parental care.

Mallards and some other ducks are dabblers, while others are divers. Divers are less buoyant, diving for fish or other prey. Diver's feet are set back on the body, and the hind toes are webbed to assist in swimming. Divers are poor walkers.

L: 23"   W: 35"

Turkey Vulture
Cathartes aura

The naked red head makes it look smaller than a similar-size hawk. All black, but the rear part of the wings look gray while soaring. Differs from hawk and eagle in flight; the vulture's wings are more V-shaped and the tips of the wings bend upward (a dihedral shape). A decomposer, the vulture cleans the environment by disposing of the bodies of deceased squirrels, skunks, raccoons, and such. A good sense of smell enables it to find food. A very acid stomach and a habit of defecating on its own feet keeps the bacterial load down. Wards off attackers by vomiting. Migration is triggered by change in day length. Hinckley, Ohio, celebrates the annual return of Turkey Vultures on Buzzard Day in February.

L: 26" W: 67"

Northern Harrier
Circus hudsonius

Once called Marsh Hawk, the Northern Harrier is easily identified by its usual behavior of soaring and circling low over fields or wetlands, showing a distinctive barred tail, white rump, and a V-shaped wing posture. Long-winged and long-tailed, its owl-like face may help in detecting sounds. Unlike other hawks, harriers show distinct sexual dimorphism. Males are gray above and lighter below with black wingtips. Females and juveniles are darker brown above with a streaked buffy underside. They may travel over 100 miles a day in search of small mammals and small birds (although they may occasionally feed on ducks). Usually seen soaring alone over wide open spaces, the Northern Harrier may share communal roosts with other harriers in the colder months of winter.

L: 18" W: 43"

Red-tailed Hawk
Buteo jamaicensis

A common hawk found throughout North America, the adult has a rusty, not red, tail. Immature birds have a banded tail until 3 years old. Body darkness varies from very light to very dark, but all Red-tailed Hawks show a darker breast band over a lighter chest. Often seen soaring over open fields or perched on fences or utility poles. Feeds mainly on rodents, such as rats and squirrels, but will also take lizards, frogs, and other birds. Builds large, open nests. Once called the chicken hawk, the Red-tailed Hawk's ecological role in pest management outweighs the minimal damage done to chicken coops. The Red-tailed Hawk, like most birds, is protected by federal law.

L: 19"   W: 49"

American Kestrel
Falco sparverius

The smallest American falcon is found throughout much of North and South America. Males have blue-gray wings with black spots and white undersides with black barring. The back is rufous and barred on the lower half. The tail is also rufous with a terminal black band. The face features two narrow, vertical black facial markings on each side. Feeds on small animals such as grasshoppers, crickets, lizards, mice, and small birds. It is often seen along roadsides and fences, hovering with quick wingbeats as it scans for prey. The kestrel attacks its prey on the ground, except for birds, which it captures in midflight.

L: 10"   W: 22"

American Coot
Fulica americana

A common denizen of watery areas, the all-black coot sports white wing edges and a white bill encircled by a black ring. Long, lobed toes help it swim. Stubby wings require long runs on water before takeoff, but often the runs are sufficient. Sometimes called white-bills or mud hens, coots lay 6–7 eggs on a mat of floating vegetation. Young swim soon after hatching and feed on the usual diet of aquatic organisms and vegetation. Vaguely resembling a duck, coots are a member of the rail family and a distant cousin of cranes. A hunted bird, it is easy to shoot but not good to eat.

L: 15"   **W: 24"**

Ocean   Freshwater   Marshes

Killdeer
Charadrius vociferus

The Killdeer's common name comes from its loud call, "kill-deer, kill-deer." Upperparts are mostly brown, head has patches of black and white, and the white breast features two black bands. Rufous tail and rump. A common shorebird found in grassy and open habitats, including grasslands, schoolyards, and golf courses. Eats various invertebrates. Simple nest is just a depression in gravel. Four spotted eggs are positioned with ends inward to prevent rolling; if one egg is destroyed, parents replace it with a rock. Young are speckled like eggs and can run quickly, although they will freeze at the approach of an intruder. To fool predators, parents will often feign injury by drooping a wing and leading the intruder away.

> **Killdeer have disruptive coloration: The two brown breast bands break up the outline of the bird, making it harder to see.**

L: 10" W: 24"

Spotted Sandpiper
Actitis macularius

The most common sandpiper in North America, the Spotted Sandpiper is also one of the most common sandpipers along rivers and streams. Its short tail, bobbing behavior, and white breast with brown spots give it away. Rarely walks but runs in short spurts; in flight, it alternates quick flapping and gliding. Lays four spotted, well-camouflaged eggs in a simple depression in gravel.

Many other sandpipers and shorebirds live along the Atlantic Coast and inland lakes. Some are difficult to identify, with small ones lumped into a nondescript category of "peeps."

Head or body bobbing is common among some birds and thought to give the bird a better perspective of its environment, much as a golfer looks at the green to line up a putt. It can also be part of courtship behavior.

L: 7"   W: 15"

Ocean

Freshwater

Common Tern
Sterna hirundo

The common tern has a white body and tail, a
gray back, a black cap, and an orange-red bill.
It flies buoyantly and nests on any flat, poorly
vegetated surface close to water. The nest may be
a bare depression in sand or gravel. Dull colors and
blotches provide camouflage for eggs. Nests in Canada,
but look for it during migration time.

L: 14"   W: 31"

Black Tern
Childonias niger

This small, black-bodied tern with gray wings, back, and tail
is also a migrant in the eastern US. It flies erratically and
inhabits freshwater ponds and marshes, building floating
nests on vegetation among rushes.

L: 9"   W: 23"

**Terns and gulls are related but have significant differences. Terns
have straight bills; long, narrow wings; fast flight; and dive for
fish. Gulls have a more rounded upper bill and the tip curves over
the lower; with a steady wingbeat and soaring flight, they rarely
dive but prefer to pick prey from the water's surface.**

Herring Gull
Larus argentatus

One of the most common gulls in North America and a resident of the eastern US and West Coast, the Herring Gull is white with a light-gray back and pink legs. Often sitting on water, gulls eat almost anything, dead or alive, and often become pests at landfills and picnics. They nest on rocky lakeshores, islands, or cliff ledges. Two or three young hatch and peck at the red spot on the parent's bill to stimulate food regurgitation. The similar, very common **Ring-billed Gull** (*Larus delawarensis*; L: 18", W: 48") is almost identical to the Herring Gull, except for its yellow legs and black ring around the bill instead of a red spot.

L: 25"   W: 58"

Mourning Dove
Zenaida macroura

The Mourning, not morning, Dove is so called because of its plaintive call—a mournful hooting. Mourning Doves are grayish brown with black splotches on rounded wings. With a pointed tail, it is smaller and faster in flight than the Rock Pigeon, with its nearly squared-off tail. In flight, it shows its white outer tail feathers. Often sits on wires and fences. Like other pigeon relatives, it lays two eggs in a saucer-shaped nest on or near the ground. The naked, helpless young are fed "pigeon milk," the sloughed-off lining of the parent's crop. Common across the continent, Mourning Doves may gather in large flocks during migration.

L: 12"   W: 18"

Rock Pigeon
Columba livia

Found all over the US, the Rock Pigeon, once called Rock Dove, is the common city pigeon found around the world. Color varies from white to black, with a mottled-gray color being most common. Has an iridescent green neck and a structure over the base of the bill called an operculum. A city bird, it eats most anything, although it prefers grains. Nesting on rooftops and window ledges that remind them of their original cliff-dwelling habitat in Europe and Asia, they might have six broods in a year. Since the time of ancient Egypt, pigeons have been bred into more than 1,000 varieties.

L: 12"   W: 28"

Barn Owl
Tyto alba

One of the most widely distributed birds in the world, the Barn Owl is a medium-size, pale tan–colored owl with long wings; a short, square tail; and a heart-shaped face. The owl's head and upper body vary between pale brown and some shade of gray. Searches at night over prairies and farms for rodents, using vision that far exceeds that of humans. Humanlike ears, hidden beneath feathers, can accurately pinpoint the source of the quietest sound. Fringed flight feathers make for quiet flight. Like other hawks and owls, eggs hatch in the order they are laid, resulting in different-aged young that compete for food in the nest. Owls regurgitate undigestible fur, feathers, and bones, producing owl pellets. Look for them under the nesting tree.

L: 16"   W: 42"

Ruby-throated Hummingbird
Archilocus colubris

The smallest bird and the only hummingbird east of the Mississippi River. Back, head, and tail are green; breast and belly are white. Male's iridescent red throat may look black in poor light. Voice is a squeaky buzz. Hummingbirds' wing structure is modified so that they can hover and fly backwards. Their high metabolism supports rapid wingbeats of 50 beats per second. A mop-like tongue and long, narrow beak allow hummingbirds to feed on nectar in various flowers. They must eat at least half their body weight in nectar every day, typically feeding five to eight times an hour. Will also catch insects in midair during breeding season when young need protein. They winter in Mexico and Central America, flying more than 500 miles nonstop across the Gulf of Mexico.

L: 3"   W: 4"

Belted Kingfisher
Megaceryle alcyon

Found throughout North America, the shaggy-headed blue-and-white kingfisher is easy to identify, although only the female has the rusty chest belt. Its long, chattering flight call is distinctive. Perches at stream sides with a good view of the water and dives for fish; may even swim a bit. Brings fishy prey back to perch to beat it on a branch before swallowing it headfirst, to avoid getting it stuck in the throat. Usually alone, except during breeding season when a pair digs a tunnel in a streambank, assisted by two toes that are fused together. Lays about seven white eggs that need no camouflage because they are far back in the burrow.

L: 13"   W: 20"

Freshwater

24

Downy Woodpecker

Dryobates pubescens

This small woodpecker has a white belly and back, white-spotted black wings, white outer tail feathers, and a black-and-white-striped head; the male has a red patch on the back of the head. Found year-round throughout the lower 48 states, except for parts of the Southeast. Downy Woodpeckers prefer open deciduous forests, although they frequent a variety of habitats, including parks and backyards. They nest in tree cavities, generally laying 3–6 eggs. The very similar but less common **Hairy Woodpecker** (*Dryobates villosus*; L: 9", W: 15") is bigger, but it's easier to distinguish them by relative bill size. The Downy's bill is about one-third of its head length, while the Hairy's bill is about half its length.

L: 7"   W:12"

Red-bellied Woodpecker
Melanerpes carolinus

A misnomer, the Red-bellied Woodpecker has only a pinkish wash on its belly but has a wide red band from its beak to its nape. The female lacks this red band. The face, throat, and underside are grayish; the back and wings are black-and-white striped; and the rump is white. Red-bellied Woodpeckers are common in woodlands, wetlands, and suburban trees in the eastern half of the US. Often seen on branches or trunks of trees, picking at the bark in search of insects. Typical woodpecker features include stiff tails to prop up their bodies against tree trunks, as well as feet with two toes facing forward and two back, which help them to hang on to bark as they navigate up and around tree trunks.

L: 9"   W: 16"

Farms, Parks, Cities   Woods   Freshwater

Northern Flicker
Colaptes auratus

This common large woodpecker is often seen probing the ground. Brown with black bars on the back and wings and a necklace-like black patch on the upper breast; the lower breast and belly are beige with numerous black spots. The white rump is conspicuous in flight. Males have a red "mustache" at the base of the beak. Because the underside of the wing can vary from red to orange to yellow (it's mainly yellow in the eastern US), the Northern Flicker was once thought to be two species.

The Northern Flicker is often seen sitting on an anthill, allowing ants to crawl over its body. The bird then grabs one ant, crushes it, and rubs it through its feathers. It may repeat this behavior several times. The ants contain chemicals, such as formic acid, that can kill insects, mites, fungi, and bacteria. Many other species of birds engage in "anting."

L: 12"   W: 20"

Eastern Phoebe
Sayornis phoebe

Named after its raspy "fee-bee" call, the Eastern Phoebe is one of the earliest songbird arrivals in the eastern US. Insectivorous like other flycatchers, it also eats berries, allowing its arrival on the summer breeding grounds before insects are abundant. Unlike most small flycatchers, it is easy to identify by its call and habit of repeatedly wagging its tail up and down. Most often found in sparse woodlands, primarily near water when nesting. During migration and in wintering in the Deep South, generally found around woodland edges and brushy fields. Many nests are built on artificial structures, especially under bridges. The shelflike nest is built of mud and strengthened with twigs, leaves, grasses, and moss.

L: 7"   W: 10"

Eastern Kingbird
Tyrannus tyrannus

This flycatcher has a bright-white chest and neck that contrasts with a black back, head, face, and tail. Tail is edged with white. Small red patch on head is rarely visible. Inhabits open areas such as meadows and prairies. Upper bill ends in small hook to help hold insects, the largest of which are beaten on a branch to kill and soften. May eat fruit as well. Aggressive, as indicated by its scientific name, the Eastern Kingbird will chase larger birds like crows and hawks, which it can do with impunity since it is more agile than soaring birds. Falcons remain a threat, though.

L: 9"   W: 15"

Red-eyed Vireo
Vireo olivaceous

Abundant during the summer months in wooded areas, the Red-eyed Vireo is one of the easiest vireos to identify due to its gray crown with distinct dark border, white stripe over the eyes, black stripe through the eyes, and, of course, red eyes. Its back is a drab olive, while its front is white. Song resembles that of the American Robin. Similar in shape to warblers, vireos are slower and more deliberate in their movements compared to the flighty warblers. Cup-shaped nest is usually placed in the fork of a tree branch and made of bark, moss, lichens, and leaves, often held together with spider silk.

L: 6"   W: 10"

American Crow
Corvus brachyrhynchos

Black with a rounded tail in flight, the crow's call is a nasal, high-pitched "caw." It's not very pleasant, but crows are songbirds with a complex voice box and have many different calls. Migrates and uses virtually all types of habitats. Winter flocks may number in the millions. Omnivorous, crows eat almost anything they can fit in their beak— insects, small birds and mammals, worms, nuts, berries, and even carrion. Although black, their feathers are iridescent and seem to change colors as light hits them from different angles.

L: 17"   W: 39"

Blue Jay
Cyanocitta cristata

An iconic bird of the eastern US, the Blue Jay is loud and colorful and has a crest, a white face with a black necklace, blue-and-white back and tail, and whitish underparts. Known for its raucous call like a squeaky water pump, it can also mimic the calls of local hawks. Feeds on berries, seeds, and nuts but will occasionally eat nestling birds. A denizen of deciduous woods, Blue Jays are also common in parks and gardens and on bird feeders. Blue Jays cache acorns for future use; because they do not retrieve all of them, the birds are responsible for the distribution of many oak trees.

L: 10"   W: 15"

Barn Swallow
Hirundo rustica

Found over most of the world, it is the only swallow with a long, forked tail. Metallic blue above with a cinnamon throat and rusty belly, it resembles the **Cliff Swallow** (*Petrochelidon pyrrhonota*; L: 6", W: 14"), which has a white neck band but lacks the long tail. Barn Swallows nest solitarily on or in artificial structures, while Cliff Swallows nest in colonies of up to several hundred birds, placing nests made of mud on or under bridges. **Bank Swallows** (*Riparia riparia*; L: 5", W: 13") and **Northern Rough-winged Swallows** (*Stelgidopteryx serripennis*; L: 5", W: 14") are both brownish, but the Bank Swallow is white underneath with a chest band; both nest in riverbank burrows. The **Tree Swallow** (*Tachycineta bicolor*; L: 6", W: 14") is bluish-green above and white below and uses tree cavities. All swallows feed on insects while in flight, aided by their streamlined body and long, pointed wings.

L: 7" W: 15"

Farms, Parks, Cities

Purple Martin
Progne subis

The largest of the swallows, the Purple Martin feeds aerially on insects, nests in cavities, and winters in South America. Despite its name, the bird is not really purple. The blackish-blue feathers are iridescent, giving them a bright-blue to navy-blue or deep-purple appearance. Almost all Purple Martins nest in specially made colonial birdhouses called "martin mansions." Native Americans would hang hollow gourds around their villages to attract martins.

> A long-standing myth is that Purple Martins eat 2,000 mosquitoes a day, hence the bird's popularity as a pest control. Research, however, has proven that Purple Martins rarely eat mosquitoes, simply because the insects contain so little nutrition and are not worth the energy to pursue.

L: 8"   W: 18"

Farms, Parks, Cities

Freshwater

Black-capped Chickadee
Poecile atricapillus

This small bird has a black cap and throat, a gray back, and a lighter gray belly; sides of the face are white. Named after its "chick-a-dee-dee-dee" call. Found all year across Canada and the northern US, where it eats insects, larvae, and eggs, along with some seeds. Nests in birdhouses or tree cavities. Solitary in the summer but flocks in the winter for protection. In winter it eats mainly seeds and frequents bird feeders. The **Carolina Chickadee** (*Poecile carolinensis*; L: 5", W: 8") is found in the southeastern US, so the two species barely overlap. Although similar, the Black-capped has a whiter nape and brighter white edges on its secondaries. The voice of the Carolina Chickadee is higher pitched and one syllable longer.

L: 5"   W: 8"

Forests

Tufted Titmouse
Baeolophus bicolor

A little gray bird with a distinct topknot, the Tufted Titmouse is common in eastern deciduous forests and a frequent visitor to feeders. Year-round residents in deciduous forests and common in parks and suburbia, Tufted Titmice flit from branch to branch looking for food, often accompanied by nuthatches, chickadees, and kinglets. In summer they mainly eat insects, such as caterpillars, beetles, ants, wasps, spiders, and slugs, but they will also eat seeds, nuts, and berries. They often hang upside down as they investigate cones, undersides of branches, and leaf clusters. Holding a large seed, such as those of sunflowers, with their foot, they hammer it open with their beaks.

L: 7"   W: 10"

Forests

White-breasted Nuthatch
Sitta carolinensis

This deciduous forest resident has a gray body; white belly, breast, and face; a black head and nape; and a chisel-like beak. Most often seen spiraling down a tree trunk one foot at a time, it probes the bark for invertebrates, seeds, and nuts. The nuthatch has a large back toe that helps it cling to bark. Reaching the bottom, it flies to the next tree and repeats. Its name comes from the habit of jamming a nut into tree bark and then "hatching" out the seed by pecking at it. Mainly coniferous forest inhabitants, the similar **Red-breasted Nuthatch** (*Sitta canadensis*; L: 4", W: 18") is smaller, with a white stripe over its eyes, a black stripe through its eyes, and reddish undersides.

L: 6" W: 11"

House Wren
Troglodytes aedon

Found in woods and cities, this small grayish-brown bird with black barring on the wings and tail has a long, thin, downcurved bill for insect eating. Often holds tail nearly vertical, especially when singing long, complex, melodious songs from exposed perches. Prefers shrubby vegetation in deciduous woods but can also be found in brushy habitats over most of the US and southern Canada. Nests in tree cavities, rock walls, and emergent vegetation. May have two or three broods with 6–8 eggs each. Parents may make as many as 600 trips a day to feed nestlings. Easily attracted to nest boxes. Aggressively defends nest. The **Carolina Wren** (*Thryothorus ludovicianus*; L: 6", W: 8"), found over much of the eastern US, is a more striking bird, with a rufous back, tan belly, and a white line over the eye. Extremely varied voice.

L: 5" W: 6"

Woods | Farms, Parks, Cities | Marshes

American Robin
Turdus migratorius

Perhaps the best-known American bird, the American Robin is identified by its black head, dark-gray back, and rufous breast and belly. Known as the harbinger of spring, robins hold winter territories even in very cold environments. They eat a varied diet while on the ground and can hear earthworms moving through their tunnels. Aggressive during nesting season and may attack its reflection in windows and car mirrors. Lays 3–5 light-blue eggs in a nest made of mud and grass. Young robins (and other songbirds) hatch in about 10 days. After another 10 days, they jump from the nest, even though they can't fly. Parents will take care of juveniles on the ground until their feathers grow enough to allow them to fly. People mistakenly think baby robins have fallen from their nest and need help. They don't.

L: 10"   W: 17"

Eastern Bluebird
Sialis sialis

The familiar Eastern Bluebird has a blue back and head, orange breast, and a white belly. A cavity nester whose holes have been disappearing because of habitat loss and the erection of plastic fences that woodpeckers cannot drill; this has instigated the popular pastime of building bluebird houses. Largely insectivorous, bluebirds are year-round residents of the southeastern US, but some migrate to northern states during breeding season. Female bluebirds tend to be polyamorous; four out of five eggs may be fathered by three or more males.

L: 7"   W:13"

Prairies

> Like many other species, bluebirds show sexual dimorphism; that is, the male and female look different. In bluebirds, it is a matter of color intensity; the female looks much duller than the male. The reason is the role of the two sexes. The male is bright to attract females and to advertise his territory to intruders. The female is dull so as not to be easily noticed by predators while raising young.

Wood Thrush
Hylocichla mustelina

This bird's rufous head and back contrasts with its white chin, chest, and belly marked by dark-brown, almost circular spots. Common thrush of deciduous forests throughout eastern US and southern Canada in summer. Winters in Mexico and Central America. Good singer with a melodious flute-like song. An omnivorous ground feeder, it eats berries, seeds, insects, and other soil invertebrates. Builds nest of mud and twigs in tree close to the ground. The similar **Hermit Thrush** (*Catharus guttatus*; L: 8", W: 12") is dull brown with a rufous tail, with the spots on its white front limited to chin and upper breast. A resident breeding bird of most of the US, it spends the winter in the southern states, down to Central America.

L: 8"   W: 13"

Gray Catbird
Dumetella carolinensis

Called catbird because, among its complex songs and mimicking of other birds, it "meows." Identified by an all-gray body with a black cap and rusty undertail feathers, the omnivorous catbird prefers low-and-dense vegetation. Its scientific name, *Dumetella,* means "small thicket." In courtship, the male catbird droops its wings and raises its tail to display a rusty patch underneath. Sitting on 5–6 eggs, the female catbird sticks her tail nearly vertical.

The mockingbird, Brown Thrasher (***Toxostoma rufum***; L: 12", W: 13"), **and the catbird are all in the same family of mimics (Mimidae). The mockingbird repeats its song phrases twice, the thrasher once, and the catbird just sings phrase after phrase.**

L: 9"   W: 11"

Farms, Parks, Cities

Northern Mockingbird
Mimus polyglottos

The robin-size Northern Mockingbird, the state bird of five states, is gray on the back and lighter in front, with white wing patches in flight. Its long tail is black in the center and dark on the sides. Mimics other birds with a varied repertoire, repeating each song twice. It will often sing all night, especially during a full moon. An omnivore, it eats mainly insects and fruits. Intelligent, mockingbirds can recognize individual humans. It prefers open areas with little vegetation. Once restricted to the southern states, mockingbirds are now found across the US and in parts of Canada. Climate change has warmed the environment, and many birds are moving their ranges northward.

L: 10"   W: 14"

European Starling
Sturnus vulgaris

Imported from Asia into New York's Central Park in the late 1800s, these starlings later spread to the Pacific Coast within a few decades and became pests in many urban areas. Starlings have a shorter tail than other all-black birds, and in fall, they have a black bill and blue-black metallic sheen on their body. In spring, the bill turns yellow and wear on the body feathers exposes numerous white spots. Gives a series of high whistles and, related to mynahs, is a good mimic. Eats insects and grains. Nests in cavities and often outcompetes native birds for nest holes.

The European Starling, House Sparrow, and Rock Pigeon are not protected by federal law, as all other birds are, because they are nonnative. The starling was imported by the American Acclimatization Society, which introduced European plants and animals into North America for economic and cultural reasons.

L: 8"   W: 16"

Cedar Waxwing
Bombycilla cedrorum

The Cedar Waxwing is a resident of the northern US, where it breeds in coniferous forests, and a winter visitor in the South. It's recognized by its silky plumage, black face mask, and permanent topknot. The tips of the secondary wing feathers appear to have been dipped in sealing wax. The greater number of red tips a bird has, the more mature it is. Most often seen in flocks in winter, when they descend on berry bushes to gorge themselves. Social creatures, a satiated bird will pass a berry onto its neighbor. Occasionally, older fermented berries will cause the birds to become tipsy and even fall from their perch. In summer, they prefer insects.

L: 7"   W: 12"

SUMMER   WINTER

Northern Cardinal
Cardinalis cardinalis

The bright-red body, black face, and topknot of the male cardinal are iconic. The female is a dull red with a green back. A common year-round resident of most of the eastern US, the Northern Cardinal is the state bird of Illinois, Indiana, Kentucky, North Carolina, Ohio, Virginia, and West Virginia. It inhabits woods, parks, and backyards, making its presence known with a loud, clear whistle. Unlike most songbirds, the female sings, and often cardinal pairs will share song phrases. The cardinal's song sounds like "cheer, cheer, cheer" or "purdy, purdy, purdy, purdy." They frequent bird feeders if sunflower seeds are offered. Typically, one out of five pairs mates for successive years, but most "divorce" between seasons.

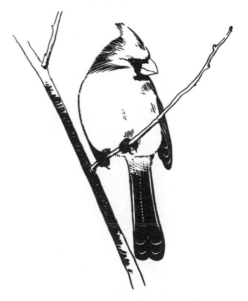

L: 9"   W: 12"

Yellow-rumped Warbler
Setophaga coronata

The most common warbler in winter. Whether in its striking summer plumage with a black chest or its brownish-gray winter outfit, it can always be identified by the yellow rump, throat, and flanks. Flits rapidly through vegetation, picking insects and larvae from twigs and leaves. They often sally out, flycatcher-like, to capture insects in midflight. Warblers have complex "warbling" songs and thin bills for reaching into bark crevices for small food items. Most warblers migrate south to spend the winter and seek invertebrates and fleshy berries. The Yellow-rumped Warbler, however, can digest wax myrtles, indigestible by most other birds, allowing it to winter farther north than any other warbler.

L: 6"   W: 9"

Warblers are small flighty birds with thin bills that sing complex, melodious songs. Adult males in breeding season are brightly colored, but in winter they're drab like the females. Feed on insects and their larvae and eggs. Build small cup-shaped nests of grass, bark strips, leaves, and moss in trees or on the ground.

Breeding in the northern US and Canada, they winter in Central and South America. About 40 species are found east of the Mississippi River. Common warblers include the **American Redstart** (black with pinkish-orange patches); **Black-and-white Warbler** (black-and-white stripes); **Yellow Warbler** (yellow with red streaked necklace); **Wilson's Warbler** (yellow with black cap); **Chestnut-sided Warbler** (yellow cap, chestnut sides); **Common Yellowthroat** (yellow with black face mask bordered by white stripe); and **Palm Warbler** (streaked breast, yellow throat, cinnamon cap).

Eastern Towhee
Pipilo maculatus

Distinctive with a black hood, red eyes, white-spotted black wings, a black tail, rufous sides, and a white belly, the Eastern Towhee is most often found in thick brush near the ground. Feeds by rummaging through leaf litter, kicking both feet backwards at the same time, looking for seeds and insects. Call sounds like "drink your tea," which someone thought sounded like "tow-hee," hence the name. They also have a "chewink" or "chirp" call. Restricted to the eastern US from Florida to Canada, northern birds migrate to the southern states in winter. They nest within 5 feet of the ground in dense brush, laying 2–5 eggs. Nests are subject to heavy predation pressure (up to 90% in some areas) from cats, skunks, raccoons, snakes, and Blue Jays.

L: 9"   W: 11"

Dark-eyed Junco
Junco hyemalis

The Dark-eyed Junco's coloration is variable, but it's basically grayish brown with a white belly and pink bill. The darker head and neck, often black, resemble an executioner's hood. Breeds in coniferous forests of the northeast and winters over much of the US. Related to similarly sized sparrows, with which it often associates, it hops around on the ground and in brush piles, looking for seeds and insects. In flight, its noticeable white outer tail feathers disappear upon landing. These white feathers may serve as communication within the flock or as a way of deterring a predator, which loses sight of the junco when the white disappears.

L: 6"   W: 9"

White-crowned Sparrow
Zonotrichia leucophrys

This large sparrow has a black-and-white-striped crown, a clear gray breast, a gray head, a pink beak, reddish-brown upperparts with dark streaks, and two white wing bars. The White-crowned Sparrow is found over much of Canada and the northern US. It winters east to Michigan and south to the Gulf Coast and Mexico. Like all sparrows, it eats seeds, grain, and insects. Its song begins with a clear whistle followed by trills. They will flock to bird feeders in winter and feed voraciously. The **White-throated Sparrow** (*Zonotrichia albicollis*; L: 8", W: 9") has a white throat, a streaked breast, and a patch of yellow in front of the eyes. Its song is also a clear whistle, followed by plaintive notes that sound like "Poor Sam Peabody."

Songbirds don't hatch knowing their song. They are born with a basic version of it but refine the song by learning from adults in the spring.

L: 7"   W: 10"

Song Sparrow
Melospiza melodia

This sparrow is identified by the brown streaks on its head, back, upper breast, and throat, as well as its white belly; it often has a spot in the middle of the breast. Distinctive song begins with three clear notes followed by a melodic series of notes, ending in a buzz. Many bird species hold a breeding territory, but the Song Sparrow also holds a winter territory. Found in shrubby areas throughout North America in wild and urban habitats, it's a year-round resident in virtually all of the US.

L: 6"   W: 8"

Farms, Parks, Cities

Woods

With about 30 sparrows and allies in the Eastern US, mostly in some shade of brown, identification can be challenging. Breast, facial, and head markings are the most important field marks, but even those can be subtle. The Song Sparrow and Savannah Sparrow (*Passerculus sandwichensis*; L: 6", W: 7") are very similar, for example, but the wider chest streaks on the Song Sparrow distinguish it.

Eastern Meadowlark
Sturnella magna

A black V on a yellow chest is identifying. Not a real lark but a blackbird, it sings like a lark with a melodious song that helps it attract a mate and declare territory in a treeless grassland. Often sings in flight to compensate for a lack of perches. Camouflaged by a brown-streaked back, it weaves its way through the grass, probing for seeds and invertebrates. Lays five eggs in a domed tunnel of grass. The range of the Eastern Meadowlark extends to the Midwest, where it overlaps with that of the **Western Meadowlark** (*Sturnella neglecta*; L: 10", W: 15"). They are virtually identical, except for their voice, which helps to keep the species separate.

L: 10"   W: 15"

Red-winged Blackbird
Agelaius phoeniceus

One of the most widespread and recognizable of all birds is the "redwing," named for the ruby-red shoulder patch of the male. Males establish their territories in a wetland by singing and raising the red patch, hoping to attract one to three sparrow-looking, streaked-brown females. Experiments in which the red shoulder patch was painted black caused the males to lose their territories to competitors. Each female builds a nest supported over water by emergent vegetation, lays four eggs, and incubates them.

L: 9"   W: 13"

Bobolink
Dolichonyx oryzivorus

Breeding in most of the northern US and southern Canada, the Bobolink migrates through the rest of the eastern states to central South America. Male is all black in front with a striking yellow nape and white wing patches. Female is yellowish brown. Its spring call "bob-o-Lincoln" provides the common name. Once called "rice bird," its scientific name means "long-clawed rice eater." Eating seeds and insects in farm fields, it may harm rice crops, but it also consumes weed seeds. Males sing a loud succession of short, clear notes in flight. Pair constructs a well-hidden hollow of grass on the ground where they lay 5–6 eggs. In the early 1900s, Bobolinks and other flocking birds were shot in large numbers and sold to New York markets for 25 cents a dozen.

L: 7"   W: 12"

Brown-headed Cowbird
Molothrus ater

L: 8" W: 12"

The male is black, except for a brown head; the female is grayish brown. The call is a distinctive but weak high-pitched whistle. Common in pastures, it feeds on insects stirred up by cattle grazing. The Brown-headed Cowbird is a nest parasite—it lays its eggs in the nests of other bird species, which hatch and raise the young cowbird. The cowbird parasitizes more than 220 bird species, laying up to 20 eggs per season. Waiting for a mother bird to leave her nest, a female cowbird can lay an egg in three seconds and be gone. The parasitized species might remove the egg, cover it up, or abandon the nest. If the cowbird egg hatches, the young bird grows quickly, demanding more food than its nestmates, often causing them to starve.

Although nest parasitism seems to be a good strategy, as it occurs in over 300 species worldwide, 97% of cowbird eggs do not become adult cowbirds.

Common Grackle
Quiscalus quiscula

Native to North America west to the Rocky Mountains in a variety of habitats, including open woodlands, farmland, and urban areas, grackles have glossy black feathers, a long tail that widens toward the tip, and a heavy, pointed bill. Male grackles are slightly larger with longer tails, and during breeding season have an iridescent head and neck feathers, which can appear green, purple, or blue; females are less iridescent and have a more brownish-black coloring. Common Grackles are omnivorous and feed on insects, fruits, seeds, and small vertebrates. They may raid bird feeders and even eat other birds' eggs and nestlings. Common Grackles are known for their loud, harsh calls, as well as their flocking behavior when they form large, noisy groups outside of breeding season.

L: 13   W: 17"

Evening Grosbeak
Coccothraustes vespertinus

This heavy-bodied bird has a short black tail, black wings, and a large light-colored bill. The adult male has a bright-yellow forehead and body and a large white patch on the wing. The female is dull gray but has the large white wing patch. Grosbeaks nest in coniferous forests but winter in large flocks in more southern areas, seeking large seeds, berries, and insects. Its heavy bill can open pine nuts and olive and cherry pits inaccessible to other birds. Its name came from French explorers who thought that it only came out in the evening.

L: 8"   W: 14"

Forests
SUMMER

Farms, Parks, Cities
WINTER

House Finch
Haemorhous mexicanus

L: 6" W: 12"

Farms, Parks, Cities

Adult male House Finches have a reddish color on their head, breast, and back, while females and juveniles are brownish gray with heavy streaking underneath. Residents throughout the US, they are nonmigratory and found in urban areas, agricultural fields, and forests. House Finches are primarily seed eaters but will eat fruits and insects, particularly during the breeding season when their young need protein. They build cup-shaped nests in trees or shrubs, and the female lays 3–6 eggs that are pale blue or white with black spots. House Finches have a variety of vocalizations, including a simple warbling song that ends in a buzz. House Finches have been affected by an outbreak of a bacterial disease that can cause conjunctivitis.

The red color of the House Finch male is maintained by ingesting fruits with carotenoid pigments. Females choose the brightest red male because it indicates he knows where the food is and can be a good provider.

American Goldfinch
Spinus tristis

American Goldfinches are found throughout much of North America, from southern Canada to northern Mexico. They are nonmigratory and can be found in a variety of habitats, including fields, gardens, and woodlands. Adult males have bright-yellow feathers with black wings and tail and a distinctive black cap—sometimes they're mistakenly called "wild canaries." Females have a more muted yellow coloration with dark wings. The American Goldfinch diet consists of seeds from weeds, grasses, and trees; they're also fond of bird feeders with thistle and sunflower seeds. They have a variety of vocalizations, including a distinctive, bubbly song that is often described as "po-ta-to-chip."

L: 5" **W: 9"**

House Sparrow
Passer domesticus

One of the most familiar of all birds, the House Sparrow, once called the English Sparrow, is found in many urban areas worldwide and is nonmigratory. Males have a black face mask, eye stripe, and nape, while females are dull brown. Introduced into New York in the 1800s, they quickly became pests, and two decades later efforts were made to exterminate them. Today, with the increasing use of pesticides and the decreasing number of insects, the House Sparrow is in decline. Not closely related to our other sparrows, it is more akin to the weaver birds of Africa.

L: 6"   W: 10"

INDEX

Other books in the pocket-size *Finder* series:

FOR US AND CANADA EAST OF THE ROCKIES

Berry Finder native plants with fleshy fruits

Bird Nest Finder aboveground nests

Fern Finder native ferns of the Midwest and Northeast

Flower Finder spring wildflowers and flower families

Life on Intertidal Rocks organisms of the North Atlantic Coast

Scat Finder mammal scat

Track Finder mammal tracks and footprints

Tree Finder native and common introduced trees

Winter Tree Finder leafless winter trees

Winter Weed Finder dry plants in winter

FOR THE PACIFIC COAST

Pacific Coast Bird Finder frequently seen birds

Pacific Coast Fish Finder marine fish of the Pacific Coast

Pacific Coast Mammal Finder mammals, their tracks, skulls, and other signs

Pacific Coast Tree Finder native trees, from Sitka to San Diego

FOR THE PACIFIC COAST (*continued*)

Pacific Intertidal Life organisms of the Pacific Coast

Redwood Region Flower Finder wildflowers of the coastal fog belt of CA

FOR ROCKY MOUNTAIN AND DESERT STATES

Desert Tree Finder desert trees of CA, AZ, and NM

Rocky Mountain Flower Finder wildflowers below tree line

Rocky Mountain Mammal Finder mammals, their tracks, skulls, and other signs

Rocky Mountain Tree Finder native Rocky Mountain trees

FOR STARGAZERS

Constellation Finder patterns in the night sky and star stories

FOR FORAGERS

Mushroom Finder fungi of North America

NATURE STUDY GUIDES are published by AdventureKEEN, 2204 1st Ave. S., Suite 102, Birmingham, AL 35233; 800-678-7006; naturestudy.com. See shop.adventurewithkeen.com for our full line of nature and outdoor activity guides by ADVENTURE PUBLICATIONS, MENASHA RIDGE PRESS, and WILDERNESS PRESS, including many guides for birding, wildflowers, rocks, and trees, plus regional and national parks, hiking, camping, backpacking, and more.